For humans, lightning is the earliest known experience with electricity.

Long ago, people were aware of electrical shocks from electric eels. Egyptian texts from 2750 BCE referred to this fish as the "Thunderer of the Nile".

Around 600 BCE, the ancient Greeks noticed a fossilized tree resin called amber would attract dried grass or feathers when rubbed with fur. Today, this is called static electricity.

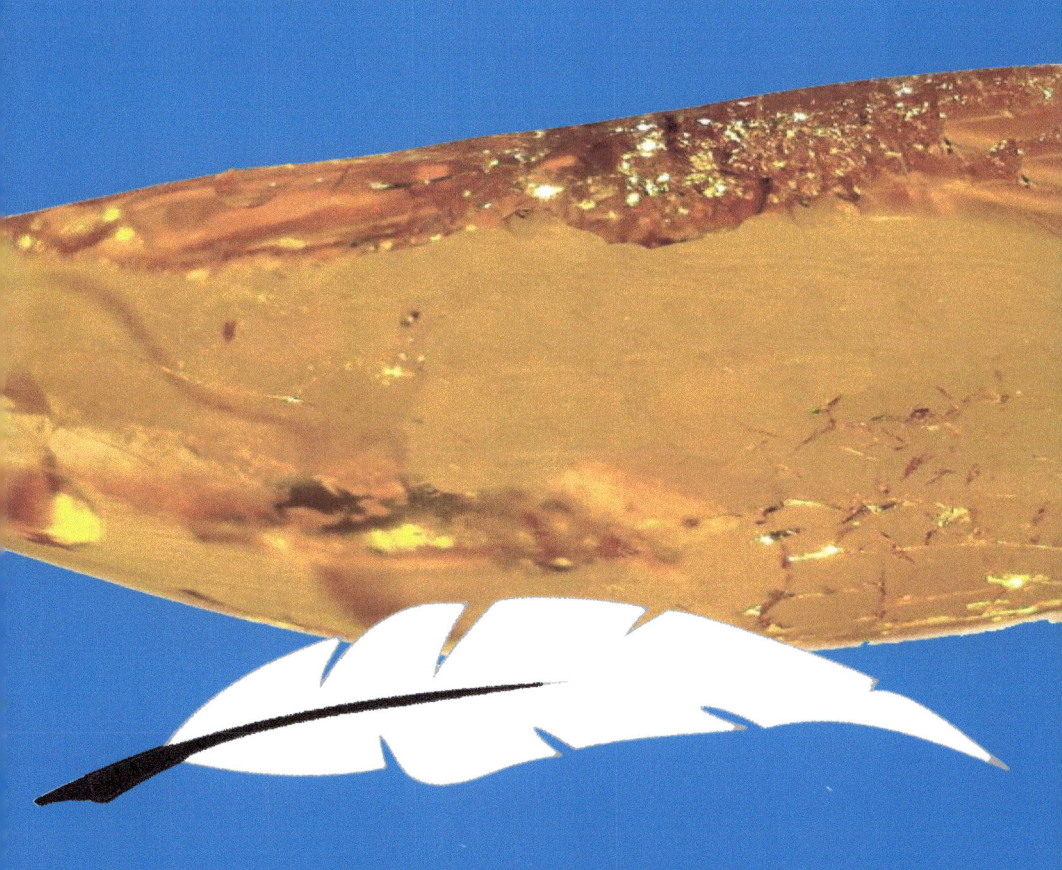

In the 1600's, the English scientist William Gilbert coined the term "electricus" from the Greek word for "amber". This word referred to those things with a property like that of amber which attract small objects after being rubbed. He demonstrated his amber experiments to queen Elizabeth.

In 1745, German experimenter Ewald G. von Kleist accidently discovered that when he touched his electric generator to a nail that was stuck into a medicine bottle through the cork, it could store an electric charge. It was called Leyden jar after the city of Leyden.

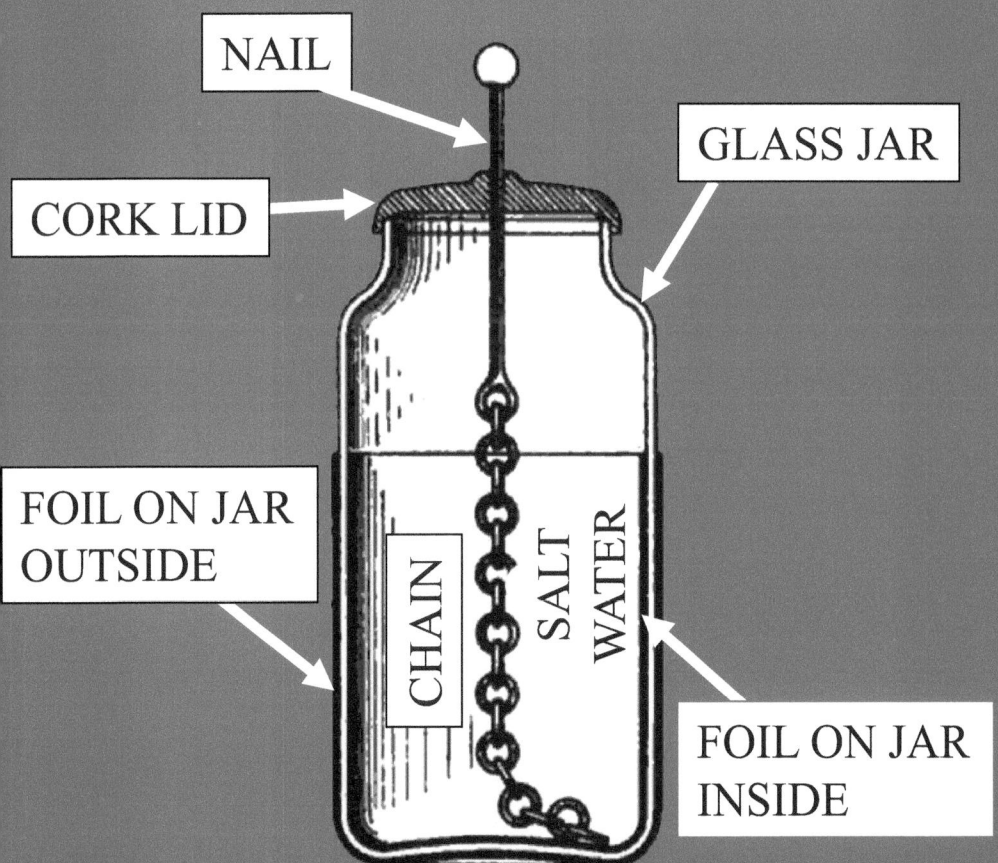

In 1752, Benjamin Franklin is believed to have attached a metal key to the bottom of a kite string and flown the kite in a lightning storm. Sparks jumped from the key to the back of his hand showing that lightning was electrical. He said the Leyden jar had both positive and negative charges.

In the 1800's, Michael Faraday was an English scientist who contributed greatly to the study of electricity. He had a huge laboratory where he did his experiments. He popularized the word ion, meaning a particle, atom, or molecule with a net electrical charge.

A neutral atom is not an ion because it contains the same number of protons as electrons. The equal positive proton charges cancel out the equal negative electron charges for a net charge of zero.

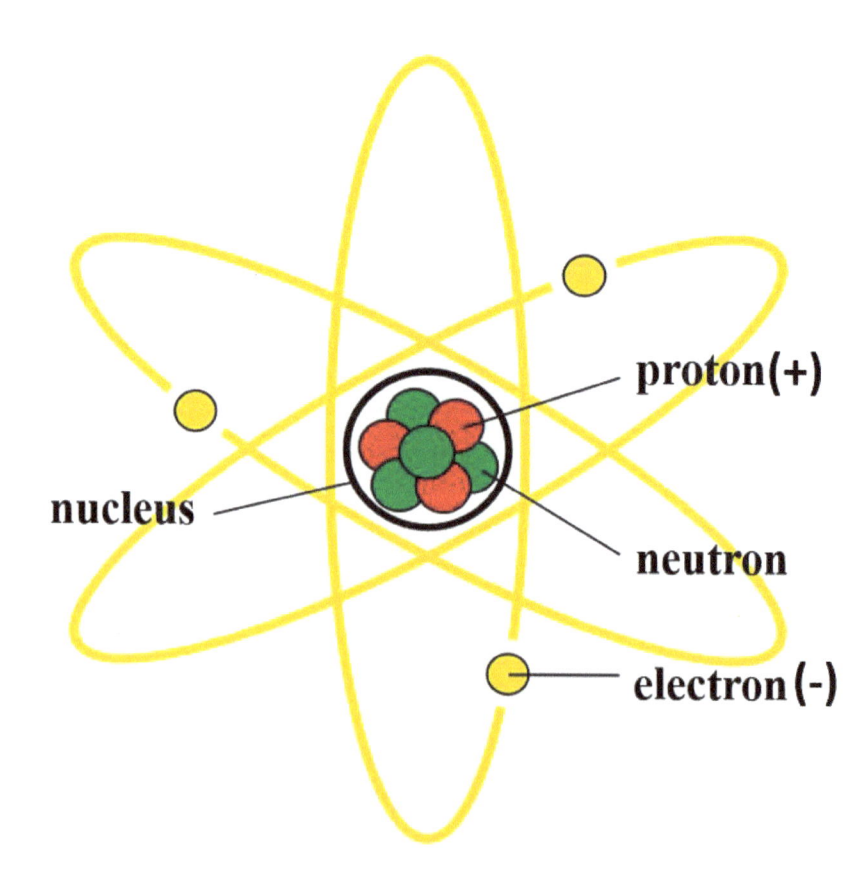

An ion is a particle, atom or molecule with a net electrical charge of negative or positive but never neutral. When a neutral atom gains one or more electrons it is a negative ion called an anion. When a neutral atom looses one or more electrons it is a positive ion called a cation.

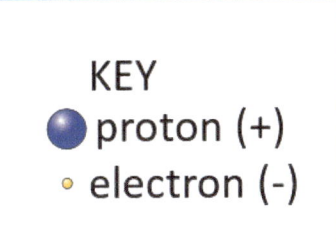

KEY
● proton (+)
○ electron (-)

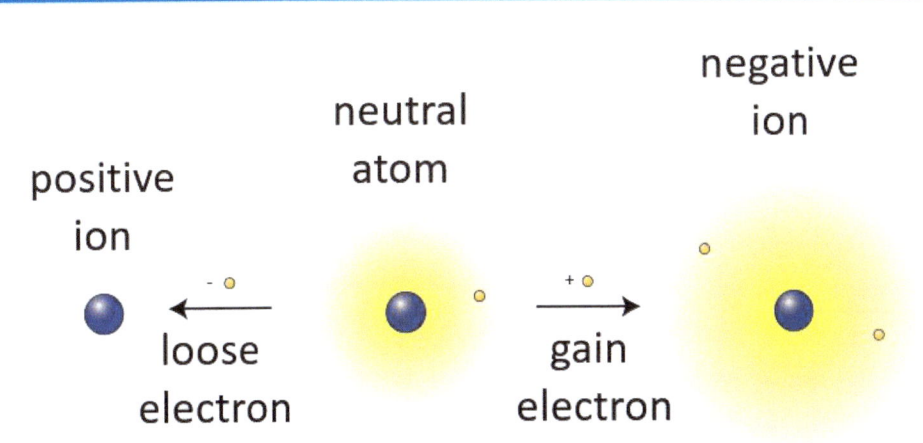

positive ion — loose electron — neutral atom — gain electron — negative ion

Ions are found everywhere n nature and in our universe. Ions are responsible for the glow of the Sun and the existence of the Earth's ionosphere that protects us from harmful radiation from outer space.

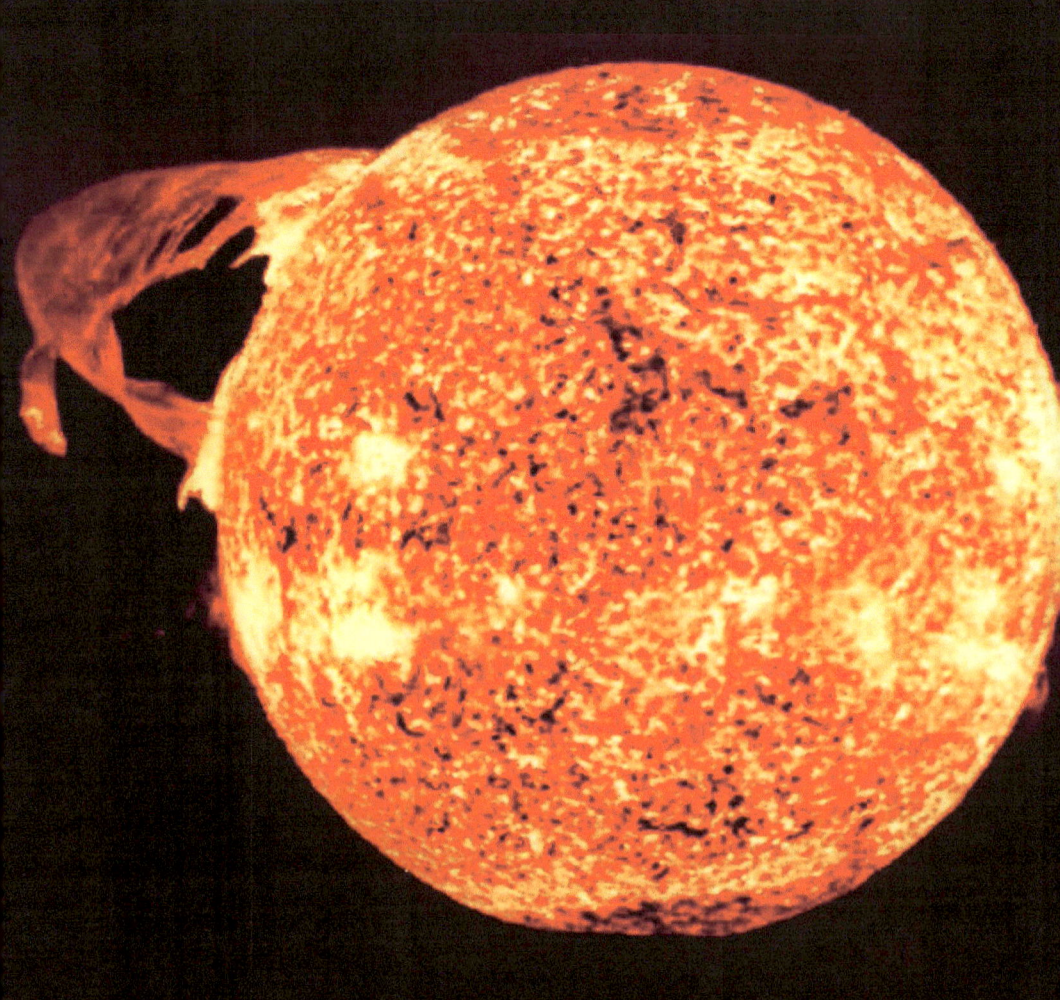

Our Earth has a plasma fountain of oxygen, helium, and hydrogen ions that gush into space from regions near the Earth's poles. This causes the aurora borealis when plasma or ion gas energy pours back into the atmosphere.

A particle accelerator uses electromagnetic fields to propel ions or charged particles at very high speeds and high energies in order to study and learn more about matter, space, and time.

Atoms in an ionic state can have a different color from neutral atoms. Light absorbed by different metal ions in gemstones gives them each a unique color.

The Periodic Table will help you learn about the patterns in chemistry and the positive and negative ions. The elements on the left side have positive ions. The elements on the right side have negative ions. The elements in the middle have positive or negative ions. Life on earth seems to exist because carbon which is found in all living things can be either a positive or negative ion.

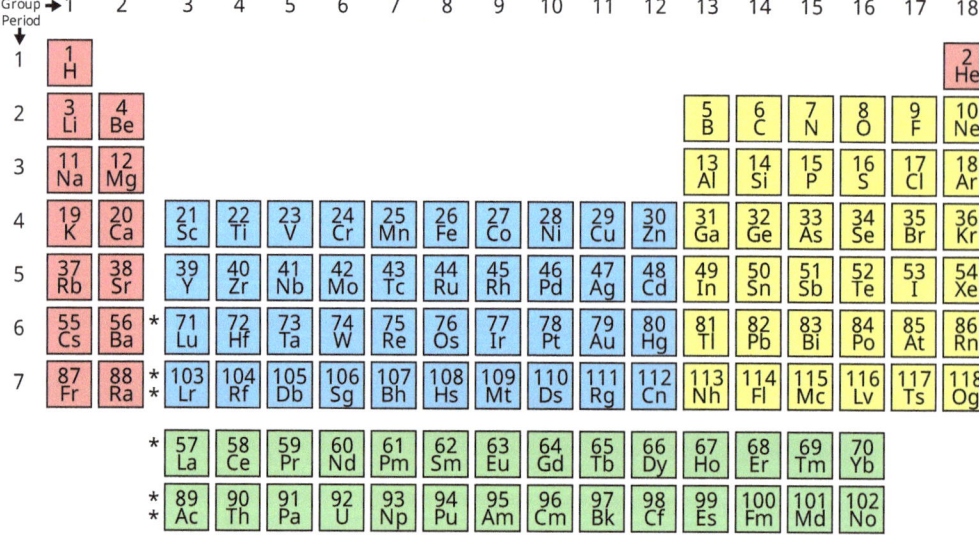

There are three ways that positive ions may be labeled in the top right corner of the element's one or two letters abbreviation, either a number with a plus sign, or one or more pluses, or else one or more pluses each inside a circle. Iron that has lost two electrons has an electrical charge of 2 plus or plus 2. This iron ion is called ferrous.

Fe^{2+} Fe^{++} $Fe^{\oplus\oplus}$

There are three ways that negative ions may be labeled in the top right corner of the element's one or two letters abbreviation, either a number with a minus sign, or one or more minuses, or else one or more minuses with each inside a circle. Oxygen that has gained two electrons has an electrical charge of 2 minus or minus 2. This oxygen ion is now called oxide.

In the Periodic Table, group 1 has one electron in the outer shell. Group 2 = 2 electrons in the outer shell. Group 3 = 3 electrons in the outer shell. Group 4 = 4 electrons in the outer shell. Group 5 = 5 electrons in the outer shell. Group 6 = 6 electrons in the outer shell. Group 7 = 7 electrons in the outer shell. Group 8 (the noble gases) is stable with 2 or 8 electrons in the outer shell.

PERIODIC TABLE ELEMENTS 1–20

HYDROGEN 1 H·							HELIUM 2 He·
LITHIUM 3 Li·	BERRYLLIUM 4 Be·	BORON 5 ·B·	CARBON 6 ·C·	NITROGEN 7 ·N:	OXYGEN 8 ·Ö:	FLOURINE 9 :F:	NEON 10 :Ne:
SODIUM 11 Na·	MAGNESIUM 12 Mg·	ALUMINUM 13 ·Al·	SILICON 14 ·Si·	PHOSPHORUS 15 ·P:	SULFUR 16 ·S:	CHLORINE 17 :Cl:	ARGON 18 :Ar:
POTASSIUM 19 K·	CALCIUM 20 Ca·						

Lewis structures also called Lewis electron dot structures (LEDS) can help you understand how elements combine and separate. The element hydrogen has one electron in its outer orbit called a valence electron. It can share electrons to form the covalent molecule H_2 which is hydrogen gas. H_2 is a very explosive, colorless gas. Hydrogen is in Group 1 and has only one valence electron. Hydrogen atoms need 2 valence electrons to have a full outer shell. Place H with one dot on the right next to H with one dot on the left to show H_2 or hydrogen gas.

The Lewis electron dot structures (LEDS) for water (H$_2$O) is the letter O with two dots on the right side. Two dots on the top. One dot on the left side and one dot on the bottom. Next to O is H with one dot pairing up with a single dot of O then H with one dot pairing up with the other single dot of O.

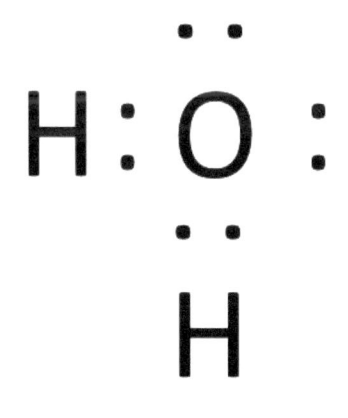

The Lewis electron dot structure for methane (CH_4) has the letter C for carbon with one dot on the top, one dot on the bottom, one dot on the left side, and one dot on the right side. There are four letters of H for hydrogen with one dot for each H that pairs with each carbon dot.

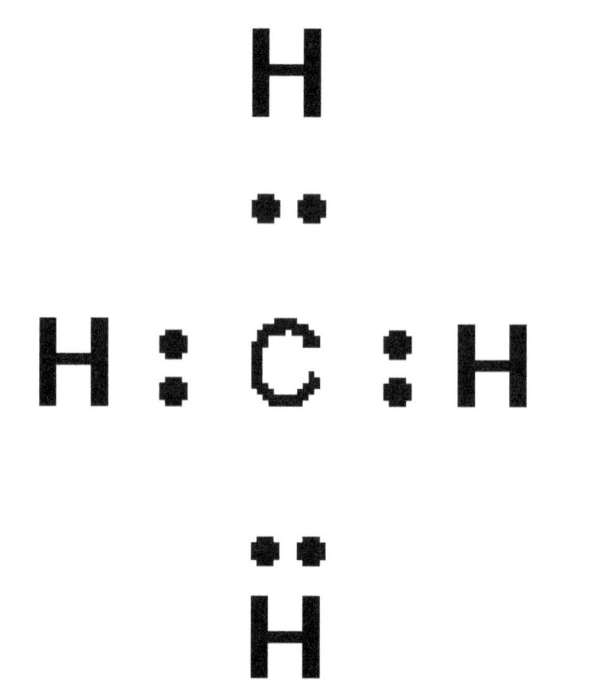

The octet rule says that some elements tend to bond in such a way that each atom has eight electrons in its outer electron shell or valence shell. In the case of hydrogen only two electrons are needed in its outer electron shell. This bonding gives the atom the same electronic configuration as a noble gas. The rule especially applies to carbon, nitrogen, oxygen, and the halogens, but also to metals such as sodium or magnesium. Carbon dioxide (CO_2) is an example of the octet rule.

Cation Examples	Anion Examples
Aluminum Al^{3+}	Azide N_3^-
Barium Ba^{2+}	Bromide Br^-
Beryllium Be^{2+}	Chloride Cl^-
Calcium Ca^{2+}	Fluoride F^-
Copper(I) Cu^+	Hydride H^-
Copper(II) Cu^{2+}	Iodide I^-
Hydrogen H^+	Nitride N^{3-}
Potassium K^+	Phosphide P^{3-}
Silver Ag^+	Oxide O^{2-}
Sodium Na^+	Sulfide S^{2-}
Zinc Zn^{2+}	Selenide Se^{2-}
Ammonium NH_4^+	Acetate CH_3COO^-
Hydronium H_3O^+	Formate $HCOO^-$

Why are ions so important? Life and our universe would not exist without ions to create an endless amount of compounds.

Dedicated to my lovely wife Sulastri and my grandchildren Mia and Kai as well as everyone who wants to learn chemistry.

For over 40 years, I have enjoyed teaching at elementary, high school and college levels.

Please visit my author page and follow to get new releases and updates at: Amazon.com/author/richlinville

Illustrations from PixaBay, Wiki, and purchased from Edu-Clips.com.

Please check out my other books at bookstores and online under the name Rich Linville.

Cold War

1945 to 1991

By Rich Linville

Wolves

by Rich Linville

If you like unicorn jokes, you might enjoy:

Unicorn Jokes for Kids

24 Unicorn Jokes with Pictures

Written by Richard Linville
Illustrated by 1EverythingNice, et al.

My Alaskan Race by Huskie Dog

From my point of view

Written by Rich Linville

My Basketball Blues
from the Basketball's Point of View
Written by Rich Linville

My Rocky Adventure!
By Rocky Magma
Written by Rich Linville

Someday I'd like to be
a rock instead of magma

Sharks

by Rich Linville

ISBN: 9798704566939

www.ingramcontent.com/pod-product-compliance
Lightning Source LLC
Chambersburg PA
CBHW040259220526
45473CB00002B/534